BE READY !!

A Citizen's Emergency Survival Guide

by

Christine Carrington

Bloomington, IN  Milton Keynes, UK

authorHOUSE®

AuthorHouse™
1663 Liberty Drive, Suite 200
Bloomington, IN 47403
www.authorhouse.com
Phone: 1-800-839-8640

AuthorHouse™ UK Ltd.
500 Avebury Boulevard
Central Milton Keynes, MK9 2BE
www.authorhouse.co.uk
Phone: 08001974150

First published by AuthorHouse 2/9/2007

ISBN: 978-1-4259-9289-7 (sc)

Printed in the United States of America
Bloomington, Indiana

This book is printed on acid-free paper.

CONTENTS

EMERGENCY PREPAREDNESS FOR YOUR FAMILY

Each year, thousands of people face emergency situations that could change their lives forever. Don't be caught off-guard. Know the hazards in your area and take the time now to prepare.

Taking steps ahead of time can help you cope better if a disaster strikes. Prepare with your family and household, and keep emergency supplies on-hand. This can help you avoid injury, help other people and minimize damage to your property.

Even though some problems cannot be prevented, you can reduce the damage by taking simple precautions, such as knowing the types of events common to your area and what time of year they are likely to strike.

Identify the risks

You can find out about the most common risks in your region by consulting the Canadian Disaster Database and the Natural Hazards of Canada map.

Consider natural disasters such as earthquakes as well as technological failures such as power outages and deliberate acts like terrorism. You may find it helpful to prepare a list of the risks you are most likely to face and think about how they might affect your family. This book addresses some of the risks that are considered to be common.

Make your own plan

Emergencies often strike too quickly to allow you to choose a shelter or pack an emergency kit, so prepare a list of what to do at home, school or work if a disaster strikes. Divide up tasks so that every member of your household participates as much as possible. Write down the details and make sure everybody has a copy. You will find the following emergency preparedness guide useful in making your plan and preparing an emergency kit.

Emergency Preparedness Guide

Know the safe places to be

Decide where to take shelter in your home during different situations such as hurricanes or earthquakes. Practice taking cover in the safe places at least once a year. Repeating this kind of safety drill and practicing exactly where to go and what to do is important for everyone but especially for children so they know what to expect and don't forget the instructions over time.

Agree on an alternate meeting place and shelter

Make sure each family member knows what to do at home, at school, at work if family members become separated or if it is impossible to get home. Decide where you will meet if you are separated during a disaster and choose a place, such as a friend's house or hotel, where you can stay for a few days in case you are evacuated. When choosing your shelter, remember that bridges may be out and roads may be blocked. Don't forget to plan for your pets -- they may not be permitted in emergency shelters so find a "pet-friendly" hotel or friend's home.

Select an out-of-the-area contact

Pick someone each member of the family can call or e-mail in case of an emergency. Be sure to choose someone who lives far enough away that they won't be affected by the same situation. This contact, ideally a relative or close family friend, should agree to pass news on to other family members if you get separated and call the contact from different places. Each member of the household should memorize this contact's phone number and address and keep the information with them at all times. If you live alone, develop an emergency plan for yourself with links to friends and neighbours. Learn about other emergency plans.

Schools

Learn about the emergency plans of your children's school or day-care centre; for example, you will need to know whether your children will be kept at school until you or a designated adult can pick them up or if they will be sent home on their own. Be sure that the school has up-to-date contact information for you. Keep in mind that the school's telephones may be overwhelmed with calls during an emergency. Find out ahead of time what type of authorization the school requires to release a child to someone else should you not be able to pick up your child yourself.

Your community

Learn about your community's emergency plans and authorities. Find out who you might need to call and what you might be asked to do in an emergency. You can contact local community offices to learn about their emergency plans. Find out where emergency shelters are located and whether there are designated emergency routes.

Identify the closest emergency services offices (e.g. fire, police, ambulance, public works such as gas and electrical utilities) and record their telephone numbers in a list near the telephone. You can find these numbers near the front of most telephone books.

High-rise buildings

If you live in an apartment building or residence, it should have its own emergency plan.

At the office

Your workplace may also have an emergency plan. Determine what your role is in the plan, what to do if an alarm sounds and how to evacuate the building safely.

Taking shelter

If you are advised by local officials to "shelter-in-place", you must remain inside your home or office and protect yourself there. The following steps will help maximize your protection:

- Close and lock all windows and exterior doors.

- Turn off all fans, heating and air-conditioning systems.

- Close the fireplace damper.

- Get your emergency kit and make sure the radio is working.

- Go to an interior room that's above ground level (if possible, one without windows). In the case of a chemical threat, an above-ground location is preferable because some chemicals are heavier than air and may seep into basements even if the windows are closed.

- Using duct or other wide tape seal all cracks around the door and any vents into the room.

- Continue to monitor your radio or television until you are told all is safe or are advised to evacuate.

Evacuating safely

If local authorities ask you to leave your home, they have a good reason to make this request and you should heed their advice immediately. Listen to your radio or television and follow the instructions of local emergency officials.

Practice your home evacuation plan with family members at least once a year.

- Wear long-sleeved shirts, long pants and sturdy shoes so you can be protected as much as possible.

- Take your emergency kit and cellular telephone if you have one. Take small valuables and papers but travel light.

- Lock your home and don't forget your key.

- Go to the designated meeting place in your family plan.

- Use travel routes specified by local authorities. Don't use shortcuts because certain areas may be dangerous or impassable.

- If you go to an evacuation centre, sign in with the registration desk so you can be contacted or reunited with your family and loved ones.

- Get in touch with your out-of-area emergency contact person (identified in your emergency plan) to let them know what has happened, that you are OK and how to contact you. Alert them to any separated family members.

- Listen to local or provincial/territorial authorities for the most accurate information about what is happening in your area. Staying tuned to local radio and television and following their instructions is your safest choice.

If you have time:

- Take your pets with you. They may not be permitted in public shelters, so follow your plan to go to a friend's home or "pet-friendly" hotel.

- Leave a note saying where you are going. Register at any local registration and inquiry centre so you can be contacted when it is safe to return home.

- If instructed to do so, shut off water, electricity and gas before leaving. Keep in mind that if you turn off the gas, a qualified technician must turn it back on when you return home. In a disaster, this might take some time.

Preparing your home

Go through your home with other household members and imagine what could happen to each part of it during a violent earthquake, hurricane or other disaster scenario.

Teach everybody in the household how to turn off the water, electricity and gas. Clearly label the on-off positions for these utilities. If your home is equipped with natural gas, tie or tape the appropriate wrench on the pipe to turn off the gas.

Discuss insurance coverage with your broker. Make sure that you have the right kind of insurance for the range of risks that may occur in your area. (See the Natural Hazards of Canada map and the Canadian Disaster Database.)

Make an inventory of the valuable things you own and keep a video or photographs of them for insurance purposes. Ensure that these and other important documents are stored safely in a waterproof area or box.

Always keep a class ABC fire extinguisher handy and make sure each member of your family knows how to use it. If you cannot extinguish a small fire with a portable fire extinguisher, or if the smoke becomes hazardous, leave the area. Close the door to contain the fire. Warn others of the danger. Call 9-1-1 and wait outside for the firefighters to arrive.

When you travel

When you go to a hotel, try to get into the habit of always keeping your footwear near the bed and clothing close by in case you have to leave in a hurry.

Consider traveling with a battery-powered radio, flashlight and extra batteries.

As soon as you arrive, identify the safest place in your hotel room to take shelter (such as under a table) as well as the locations of the emergency exits. Remember to read all safety materials provided by the hotel.

Emergency kits and Supplies / Essentials

- Household emergency kit

- Vehicle emergency kit

- Workplace emergency kit

You could be anywhere when a disaster strikes but it is likely that you will be at home, at work or in your car. Having supplies on hand can help you survive the emergency and, if necessary, make you self-sufficient for three days or longer.

Emergency kits should be portable so put supplies in easy-to-carry containers (such as a duffel bag or small plastic bin) ready to take with you. Store the kit in an easily accessible location such as a closet shelf on the main floor. You could also consider preparing a personalized kit in a backpack for each member of the family, ready to go in case you need to evacuate.

It is a good idea to have additional emergency supplies in your tool shed or garage in case you have to evacuate your home and can't go back inside.

Household emergency kit

Use the subsequent lists to assemble your emergency kit:

- Medical

- Food

- Water

- Shelter

- Clothing

- Toiletries

- Tools

- Communications

- Weather

- Important documents

- Additional items

Medical

Your kit should contain a complete set of first aid supplies, including:

- First aid manual

- Bandages

- Adhesive tape

- Antibiotic ointment

- Antiseptic towelettes

- Over-the-counter medications such as pain relievers

- Safety pins

- Cleansing agent or soap

- Cold pack

- Eyewash solution

- Cotton swabs

- Disposable gloves and face shield

- Gauze pads

- Include a few weeks supply of any essential regular medication such as insulin (check expiry dates regularly).

- Make a list of the prescription medications taken by each family member as well as the style and serial number of medical devices such as pacemakers, hearing aids and eyeglass prescriptions. Also include the names of family physicians and specialists.

- Add any "special needs" items for members of your household such as contact lens liquids, infant formula, diapers, items for people with disabilities, denture needs, etc.

- Masks

- Bioharzardous disposal bags

First aid training - Take a first aid course that includes cardiopulmonary resuscitation (CPR).

Food

Have at least a three-day supply of non-perishable food and water on hand. Remember to replace the unused canned and dry food in your kit every year. Choose compact, lightweight, ready-to-eat foods that your family likes and that don't need refrigeration, cooking or preparation. If food must be cooked, include a camp stove and fuel.

Never use charcoal or gas barbeques, camping heating equipment or home generators indoors.

Examples of non-perishable foods:

- Ready-to-eat canned food such as stews, baked beans, pasta, meat, fish, poultry, vegetables and fruits

- Cereal

- Trail mix

- Oatmeal cookies

- Crackers or biscuits

- Candies

- Freeze-dried foods

- Canned milk, juices, instant coffee and tea

- Staples such as sugar, salt and pepper, powdered coffee whitener, honey, peanut butter, syrup and jam

- Special dietary requirements such as baby food and formula

Remember to keep eating utensils, disposable cups and plates, a manual can-opener and a bottle opener with the food.

If you have pets, don't forget to store food for them.

Water

Store at least four litres of water per person per day - for drinking, food preparation, hygiene and dish washing. Store water in a cool dark place. If you store tap water, use plastic bottles that have been washed, disinfected and are easy to carry. Record the date that you bottled the water on the label. Replace stored tap water every six months.

If you have pets, don't forget to store approximately 30 milliliters of water per kilogram of the animal's weight per day. For example an average cat or small dog would require at least 1/5 of a litre (or half a cup) of water per day.

Keep a supply of water-purifying tablets and non-perfumed chlorine bleach in your emergency preparedness kit.

Three ways to purify water:

1. Boil it for at least 10 minutes.

2. Add water-purification tablets.

3. Add one drop liquid chlorine bleach per litre of water (or three drops for cloudy water). Stir and let sit for 30 minutes before drinking. The water should have a slight chlorine smell.

Shelter

- Small tent, plastic sheet or tarpaulin

- One sleeping bag or two warm blankets per person

- Candles and matches or lighter in waterproof containers

- Kerosene lantern and fuel

Clothing

It's a good idea to put aside one change of indoor and outdoor clothing and footwear per person. This could include:

- Sweaters

- Woolen socks

- Hat, scarf, mittens

- Underwear

- Rain gear

- Sturdy waterproof shoes or boots (the shoes should be heavy enough to protect feet from broken glass and other debris)

- Heavy work gloves

Toiletries

- Toilet paper

- Sanitary napkins/tampons

- Toothbrush, toothpaste

- Soap

- Towels

Tools

- Basic tools (hammer, pliers/wrench, screwdriver set, pry bar, assortment of fasteners, work gloves)

- A spare pry bar should also be stored outside in a tool shed or garage

- Wrench (crescent or pipe) in case you have to turn off the natural gas

- Duct tape

- Rope

- Small shovel

- Pocket knife

Communications

- Whistle (three short blasts is the recognized signal for help)

- Flashlight (one per person) and batteries. It is also a good idea to keep a flashlight near your bed, in your car and at work. Have spare batteries in each location.

- Pen, marker, notepad

- Battery-powered, solar or hand-cranked AM/FM radio or television (and extra batteries)

- List of the frequencies of local radio stations

Weather Information

The Meteorological Service of Canada which is part of Environment Canada monitors the weather 24 hours a day, seven days a week and issues watches, advisories and warnings through:

- Radio and television stations

- Internet at www.weatheroffice.ec.gc.ca

- Weatheradio - a network of transmitters across the country broadcast over special frequencies 24 hours a day

- Weather Line - a network of automated telephone answering devices that provide recorded information (look in the blue pages of your telephone book under "Weather" for the telephone number)

Check and replace batteries yearly to keep them fresh.

Important documents and Money

Remember to store important family documents and records in waterproof containers and make sure you know where they are. It is not a good idea to store them in the basement if there is a risk of flooding.

- Powers of attorney

- Will

- Insurance policies and life insurance beneficiary designations

- Contracts

- Deeds

- Stocks and bonds

- Family records (birth and marriage certificates, etc)

- Passports

- Social insurance cards

- Health cards and immunization records

- Licenses

- Savings and chequing account numbers

- Credit card account numbers and contact information for companies

- Important telephone numbers

- Printed copies of emergency checklists and instructions, such as these. You may not remember all of the instructions during an actual emergency and you may not have access to the Internet.

Money

Bank machines, credit cards and other forms of electronic commerce may not be available during an emergency. Keep some cash or travellers' cheques with your other emergency supplies and provisions.

Copies of essential documents should also be kept in a safe location outside your home such as a safety deposit box or the home of a relative who lives out of town.

Pet Preparedness

Preparing ahead of time and acting quickly are the best ways to keep your animals out of danger. They depend on you to be prepared in the event of a disaster situation. Your pet emergency kit should include a bowl, water, food, pet medication, travel cage or kennel, leash, blanket for bedding, plastic bags and paper towels for disposing of waste, immunization records, pet medical history and a favorite toy.

Vehicle emergency kit

Keep supplies in a separate pack (such as a tote bag) in case you are "on the road" during an emergency. Make a pack for each vehicle in your household.

- Booster cables, tools and tow chains

- Shovel and sand, kitty litter or other traction aids

- Ice scraper and brush

- Bottled water - at least four litres

- Canned food and opener, dried fruit, cookies and crackers

- Outdoor clothing, footwear and a backpack

- Sleeping bag(s) or emergency thermal blankets

- First aid kit and manual

- Compass

- Flashlight and spare batteries

- Waterproof matches, "survival" candle in a deep can (to warm hands, heat a drink or use as an emergency light)

- Cloths or roll of paper towel, toilet paper, moist towelettes and small plastic bags

- Cash and coins

- Map of the region where you live

- Pen/pencil and paper

- Playing cards and colouring books for children

- Warning lights or road flares, axe or hatchet, fire extinguisher, methyl hydrate (for fuel line and windshield de-icing), seatbelt cutter

- If you do not already have a cellular telephone, and if the cellular network works in your area, you may want to consider having one with you in your car for emergencies

- Extra windshield washer fluid and antifreeze

Remember to keep your gas tank at least half full all year round and nearly full in the winter. Gas pumps are likely to be unusable after a major disaster like an earthquake.

Workplace emergency kit

Keep the following items in a pack in your workplace in case you have to walk home or to safety:

- Gloves, walking shoes and outdoor clothing

- Emergency silver foil ("space") blanket

- Flashlight

- Radio and batteries (stored separately in waterproof bags)

- Whistle (three short blasts is the recognized signal for help)

- Bottled water

- Dried fruit and nuts, high-energy food bars

- Small up-to-date photos of family and loved ones for identification

- Paper with your name, home address and any special medical conditions

Additional items

- Gasoline-powered generator and appropriately rated extension cord

- Camp stove and fuel

- Portable toilet

- Extra car and house keys

- Playing cards, games, toys

- Books appropriate to family interests/ages

- Colouring books, drawing paper and crayons

- Small up-to-date photos of family and loved ones for identification

- Moist towelettes

- Facial tissues

- Comb or hairbrush

- Shampoo, deodorant

- Liquid detergent

- Insect repellent

- Plastic garbage bags and ties

- Chlorine bleach and liquid disinfectant

WHAT TO DO WHEN DISASTER STRIKES IN YOUR COMMUNITY

Disasters come in all shapes and sizes.

They strike any time, any place. They have no conscience. They show no mercy.

Most people in Canada don't think of disasters as happening here but they do. From tornadoes and floods to house fires. In recent years, disasters have driven more than four million Canadians from their homes and caused billions of dollars in damage. Payments by governments and insurers have been doubling every five to 10 years.

Though the type and magnitude of disasters vary greatly, one thing remains the same:

Disasters cause great suffering. Survivors almost always need help to meet immediate day-to-day basic needs, and often require longer-term help to rebuild their lives and deal with the trauma.

The following pages provide information on how to deal with a variety of natural and manmade disasters.

BLACKOUTS

Safety Tips

- Only use a flashlight for emergency lighting. Do NOT use candles!

- Turn off all electrical equipment that were in use when the power went out.

- Only open the freezer or refrigerator if you have to.

- Generators should not be run inside the home or in the garage.

- If you use a generator, connect the equipment you want to power directly to the outlets on the generator. Do not connect a generator to a home's electrical system.

- Listen to local radio and television for updated information.

What to do During a Blackout

- Don't Panic

- Turn Off Electrical Equipment

- Check Your Stove / Oven - Turn off all burners and your oven too, even if it's a gas range. Don't use your gas range to warm the house.

- Turn Off the Lights - Turn off all but one light (so you'll know when power is restored).

- Keep Refrigerator and Freezer Closed - Open the doors as little as possible to keep foods cold longer.

- Limit Phone Usage - Depending on how widespread the power outage is, phone lines (including cell phone systems) may be overloaded. Use the phone to report an outage and then for emergencies only.

- Stay Home

- Stay Warm / Stay Cool

After a Blackout

- Turn Off Your Generator

- Turn on Appliances Slowly

- Check Your Security System

- Check on Neighbors

- Inspect Your Food

- Replenish Emergency Kit

- Examine Your Trees

- Reset Televisions and Clocks

BOMB THREATS AND EXPOSURE TO CHEMICAL AGENTS

If it appears that chemical agents are involved, leave the emergency scene as quickly as possible and seek help. Local authorities are better equipped to address and contain this type of emergency. People who may have come into contact with a biological or chemical agent may need to go through a decontamination procedure before receiving medical attention.

Listen to the advice of local officials on the radio or television to determine what steps you can take to protect yourself and your family. Since emergency services will likely be overwhelmed, only call 9-1-1 about life-threatening emergencies.

Bomb threats

If you receive a bomb threat, stay calm and try to get as much information as possible. Although this might be difficult, try to note any unique features about the voice and any background sounds you hear over the telephone. Keep the caller on the line as long as possible and take detailed notes about what is said.

Try to note the following:

- If the speaker is male or female

- If the speaker has a distinctive accent

- If the voice is disguised, muffled or strange-sounding

- If the voice is shrill or deep

- Any background noises (e.g. traffic, bus passing, bell ringing, fax or printer sounds)

- Any indoor vs. outdoor sounds

Call the police and building management immediately afterwards.

After you have been notified of a bomb threat, do not touch any suspicious package. Leave the area where it was found. Notify the police immediately.

After evacuating a building, avoid standing in front of windows or other potentially hazardous areas. Do not block the sidewalk or street. It will need to be kept clear for emergency officials.

In the case of an explosion, get out of the building as quickly and calmly as possible. If items are falling off bookshelves or from the ceiling, get under a sturdy table or desk until the situation has stabilized enough for your safe passage. Ensure your own safety before trying to help others.

Making a bomb threat is a criminal offence. Do not try to guess whether the threat is real or a hoax. Call the police.

Suspicious packages

Suspicious packages could be delivered to your home or workplace, therefore it is good practice to be vigilant and know what to do. You know what kind of mail and packages you usually receive. Look for things that are out of the ordinary, such as unexpected mail from a foreign country. The following might help in identifying a suspicious package:

- Unfamiliar return address or none at all

- Strange odour or noise

- Protruding wires

- Excessive postage

- Misspelled words

- Addressed to a business title only (e.g. President)

- Restrictive markings (e.g. Do not X-ray)

- Badly typed or written

- Rigid or bulky letters

- Lopsided or uneven

- Excessive wrapping, tape or string

- Oily stains, discolouration or crystallization on wrapping

- Leaking

The contents of a letter or package may cause concern if:

- You see powder or a liquid

- It contains a threatening note

- It contains an object that you did not expect to receive or cannot identify

If you are worried about a package or letter you have received:

- Do not handle, shake, smell or taste it

- Leave the letter or package where it is

- Get everyone out of the room and close the door

- Call 9-1-1 (or the emergency response number in your area)

- Wash your hands with soap and water

- If applicable, alert building security or the superintendent

- Wait in a safe place until the police or fire response teams arrive

If you have opened a suspicious package:

- Leave the package where it is

- Remove any clothing that has powder or liquid on it and seal it in a plastic bag

- Get everyone out of the room and close the door

- Wash your hands or shower with soap and water

- Call 9-1-1 (or the emergency response number in your area)

- If applicable, alert building security or the superintendent

- Wait in a safe place until the police or fire response teams arrive

The police, other emergency workers and public health authorities will give advice about what to do next.

Chemical releases

Hazardous chemicals can be released by accident or through a deliberate act of criminal intent. In either case, it is important to listen to the directions of emergency responders.

Chemicals that could be used by terrorists vary from warfare agents to toxic chemicals commonly used by industry. People exposed to these chemicals could suffer injury, disease or death.

Emergency responders are trained to identify hazards and provide appropriate guidance to the public. In some situations, you should seal yourself inside the building you are in while other times you may be instructed to go to higher elevations or evacuate the area.

If you suspect that a chemical substance has been released in a closed area such as a subway or building, try to avoid breathing any of the fumes and evacuate as quickly as possible. Immediately contact the closest police, fire and ambulance services. Decontamination might be required before you can receive medical attention. Heed advice from local officials.

When an accidental chemical spill occurs, an evacuation of nearby communities is often ordered as a precautionary measure. Stay away from the accident. Listen to the radio for instructions from emergency responders in your area.

Exposure to a chemical substance may require quarantine and the attention of medical authorities. Because the type of chemical may not be known right away, treatment is based on symptoms. Keep track of things like breathing and heart rate, perspiration, dizziness, skin tone, deliriousness. Tell medical personnel and public health agencies about these or any other symptoms.

Biological agents

Biological agents are bacteria or viruses that can be deliberately dispersed to cause disease and/or death.

Anyone exposed to a biological agent should obtain immediate medical attention. If you experience sustained or unusual symptoms, seek medical treatment right away. If you have been exposed (or think you might have been exposed) to a biological agent but are not ill, you should still contact the public health authorities as quickly as possible. Officials will assess and manage the risks for anyone who may have been exposed to a dangerous substance. If need be, post-exposure treatment with antibiotics might be recommended by health officials.

Nuclear emergencies

The likelihood of a nuclear or radiological incident is remote because of the stringent controls in place for the movement and use of radioactive materials. All levels of government and the operators of nuclear facilities have emergency plans that are ready to be implemented at a moment's notice.

A nuclear emergency could be declared if there is an accident or an intentional release (or threat of intentional release) of potentially harmful radioactive materials. In either situation, the risks to health result from exposure to radiation.

As with any emergency, remain calm. Officials will quickly determine the degree of risk radiation and take immediate measures to limit the dangers of exposure.

Depending on the incident and risk to health, you could be visited by emergency services personnel who would advise you about what to do.

In chemical, biological or nuclear emergencies, it is important to follow the instructions of local officials and emergency responders for possible evacuation instructions.

During a nuclear emergency, you may be told to minimize the amount of outside air entering your home. If so, immediately close doors and windows then turn off air exchangers and heat-recovery units. If you were outside around the time of a nuclear emergency, remove your clothes as soon as possible and seal them in a plastic bag. Rinse your hair and body in the shower and then put on clean clothes from a closed drawer or closet. Find your emergency kit, turn off appliances and stay indoors until advised otherwise.

EARTHQUAKES

An earthquake happens when the earth's crust moves. You can then feel the ground shaking. Vibrations originate from a point called the epicenter.

An Earthquake's strength varies a lot for each event. Sometimes they are hardly felt and other times, they are strong enough to cause fixed objects to fall and houses to collapse! Earthquakes can also cause tidal waves, also known as tsunamis.

It is possible to measure the strength of an earthquake with the Richter scale of a seismograph. The Richter scale was created by an American seismologist named Charles F. Richter.

The Richter scale ranges from 0 (no earthquake) to 9 (largest earthquake ever recorded). For each level, the earthquake is ten times stronger. For example, an earthquake in the range of 3 on the Richter scale is ten times stronger than an earthquake in the range of 2 on the same scale.

Richter Scale

M1   Usually not felt
M2   Usually not felt
M3   Often felt, no damage
M4   Often felt, no damage
M5   Moderate earthquake. Easily felt. Minor damages to buildings near the epicenter

M6   Strong earthquake. Damages to weak buildings located several kilometers from the epicenter

M7   Major earthquake. Many damages to buildings located hundreds of kilometres from the epicenter

M8   Giant earthquake. Heavy property damages and many dead and hurt persons on hundreds of kilometers

M9   Super earthquake. Very rare. It destroys nearly everything in the region of the epicentre on thousands of kilometres

Know the facts.....

- There is no earthquake season. They can occur at any time of the day, the night or the year.

- Scientists studying earthquakes are called seismologists.

- The device used to measure the strength of an earthquake is called a seismograph.

- The last earthquake with maximum strength occurred in 1663 in the Charlevoix region.

Tectonic Plates

Have you ever noticed that South America and Africa could fit together like the pieces of a puzzle. That is because million years ago, before the appearance of man on Earth, the continents were not arranged as they are today. Nearly all continents were touching one another. But since then, continents have drifted apart.

That is because there are tectonic plates under continents and seas. What are tectonic plates? They are enormous plates of ground moving very slowly: some rise, some sink and some drift to the side. Continents rest on those plates.

Since all tectonic plates do not move in the same direction, they hit one another. These hits are called "earthquakes"!

What to do during an earthquake

1. If you are indoors, duck or drop down to the floor. Take cover under a sturdy desk, table or other furniture. Hold on to it and be prepared to move with it. Hold the position until the ground stops shaking and it is safe to move. Stay clear of windows, fireplaces, woodstoves, and heavy furniture or appliances that may fall over. Stay inside to avoid being injured by falling glass or building parts. If you are in a crowded area, take cover where you are. Stay calm and encourage others to do likewise.

2. If you are outside, get into the open, away from buildings and power lines.

3. If you are driving, stop if it is safe, but stay inside your car. Stay away from bridges, overpasses and tunnels. Move your car as far out of the normal traffic pattern as possible. If possible, avoid stopping under trees, light posts, power lines, or signs.

4. If you are in a mountainous area, or near unstable slopes or cliffs, be alert for falling rock and other debris that could be loosened by the **earthquake.**

5. If you are at the beach, move quickly to higher ground or several hundred yards inland.

Floods

Floods happen when a watercourse overflows. They could be caused by rain, thaws or ice jams.

Heavy Rain

When it rains, water falling on the ground penetrates it. But when the ground already contains a lot of water, the falling rain cannot seep through it. The water then starts flowing toward a watercourse.

When it rains a lot on a region, all water flows toward the river, but the river cannot always contain it; therefore, the river overflows.

Thaw

When spring arrives, snow and ice start to melt and the water flows toward the river. If the temperature rises too rapidly, a lot of water is produced in a short time; the river overflows and a flood happens.

Ice Jam

In spring, rivers thaw out and big chunks of ice detach themselves and drift on the river. Sometimes these chunks get jammed against a bridge or rocks. Consequently, these chunks of ice form a kind of dam.

When water cannot flow through this dam, it tries to find an alternate course to reach the river and sometimes this new course is a field, a road or someone's basement.

The *Waban-Aki* is a hovercraft used by the Canadian Coast Guard for ice breaking operations, which reduce flood risks. Power shovels are also used to break ice. They are known as "frogs"!

A flood is one of the most frequent disasters. All regions having a watercourse flowing through them could be affected. Is there a watercourse near your house? Find out if your municipality has a flood emergency plan.

What to Do During a Flood

Bring upstairs all objects from the basement that could be damaged by water. If you are asked to evacuate, do so immediately! The water level can rise faster than you think. Do not forget your emergency kit.

In all Canadian history, the worst flood happened in the Saguenay region in 1996. In two days, the quantity of water that fell was the same as the quantity flowing through the Niagara falls in four weeks! This flood was called the deluge.

Hurricanes

It is possible to reduce the impacts that a hurricane can have on you if you know your vulnerability. Once you have assessed your vulnerability, you can take appropriate actions, including having a basic plan of preparedness.

A hurricane preparedness plan includes three basic things that are important in the threat of any severe weather event, and not just for hurricanes:

1. Maintaining a disaster or emergency supply kit;

2. Securing your home and property;

3. Having a safe place to go in the event of evacuation or prolonged utility outage.

Hurricane victims are most at risk by storm surges (rising water) and high winds.

Hurricane protection involves accurate tracking and protection from blowing winds and rising waters. The storm surge (or increase in water level as much as 30 feet) can flood entire communities because of the bulge of water raised up by the eye. In such cases, evacuation is often the only method to save lives.

Pandemics

A pandemic is an epidemic occurring over a very wide area, crossing international boundaries and usually affecting a large number of people. A global epidemic.

During every flu season, even when there is no pandemic, you should:

- If eligible, vaccinate yourself and members of your family against influenza. Regional health authorities provide flu immunization programs every year.

- Cover your mouth when coughing or sneezing, and immediately wash your hands. This helps prevent the spread of infectious diseases, including the common cold and the flu.

If you develop the flu, follow general guidelines to take care of yourself:

- Rest.

- Drink plenty of fluids.

- Take acetaminophen for fever or pain.

- Wash your hands to prevent further spread of the disease.

Can a pandemic be averted ?

No one knows for sure. Influenza viruses are highly unstable and difficult to predict.

However, health authorities such as the World Health Organization remain optimistic that if the right actions are taken quickly, an influenza pandemic can be averted.

How will I know what to do in a pandemic?

There would be public announcements on TV, the radio and through other media channels, and regular updates using a variety of communications channels.

What should people do at home to deal with an influenza pandemic?

- Be alert to information on radio and newspapers. Health authorities will advise you about the availability of immunization and any steps you can take to avoid disease.

- Health professionals will provide care to the very ill and provide information on self-care or caring for family members at home.

Please

- Stay home if you are sick and keep away from other people -- avoid visitors and visiting others.

- Wash and dry your hands before handling food, after coughing, sneezing, using the bathroom, wiping or nose-blowing (whether your nose or your child's), and when looking after sick people.

- Keep coughs and sneezes covered. Tissues are best. Put the tissue in a rubbish bin.

- Give plenty to drink to people who have a fever and/or diarrhea.

- Include paracetamol (for fever) in your home emergency survival kit.

Terrorist Attacks

The threat of a terrorist incident is higher than ever before, No state or individual is immune. September 11 has confirmed the importance that preparation for terrorism should be built on systems of thinking and planning ahead. The responsibility to meet such attacks lies with all of us. The following information is intended to raise awareness of citizens of the need to plan for potential incidents and enhance their capacity to effectively manage potential risks to their environments. Emergency preparedness is a shared responsibility.

Prior to an Attack

Prepare for the possibility of a terrorist incident in your area, stay informed. Adapt, as appropriate, the same techniques used to prepare for tornadoes, fires, and other emergencies. Be prepared and observe your environment. Terrorists most often strike with little or no warning. Use caution when you travel. Locate stairways and emergency exits and develop plans for evacuating buildings, subways, and crowded public areas. Develop a Family Emergency Plan. Assemble and maintain an Emergency Supply Kit.

Examples of Common Terrorist Targets

Airports
Government buildings
Hospitals

Tourist attractions
Transit systems
Military bases
Diplomatic missions
Arenas, stadiums
Educational institutions
Communications networks
Utilities, power plants

In the Event of an Attack

EXPLOSION:

Remain calm.

- If objects begin to fall, take cover under a desk or sturdy table.

- Exit the building as quickly as possible.

- If trapped in debris tap on a pipe or wall so that rescuers can hear where you are.

- If possible, use a flashlight or whistle to signal rescuers regarding your location.

- Cover your mouth with a handkerchief or clothing.

- Stay in your area so that you don't kick up dust.

FIRE:

- Stay low to the floor at all times and exit the building as quickly as possible.

- Use a wet cloth to cover your nose and mouth.

- Use the back of your hand to feel closed doors. If the door is not hot, brace yourself against the door and open it slowly. Do not open the door if it is hot. Seek another escape route.

- Use appropriate fire exits and stairs, not elevators.

RADIATION:

Highest-risk areas are those in which buildings are likely to be destroyed by a blast or fire, or where a person would be in the open for the first two weeks.

Shielding: Place the most heavy, dense materials available between you and the source of the radiation.

Distance: The more distance between you and the source of the radiation, the less radiation you will receive.

Time: Most radioactivity loses its strength fairly quickly. Limiting the time spent near the source of radiation reduces the amount of radiation exposure you will receive.

PROTECTIVE ACTIONS

Shelter-in-Place or Evacuation:

Shelter-in-place means to stay indoors. If shelter-in-place is recommended, move all people and pets inside. Local officials will provide instructions on necessary actions. These can include:

- Closing all windows and doors.

- Ensuring your emergency supply kit is with you.

- Turning off air-conditioning, ventilation systems.

- Closing all fireplace dampers.

- Taping around doors, windows, exhaust fans, or vents.

- Wetting towels and placing them at all cracks under doors.

- Staying away from windows.

- Staying indoors and listening to emergency broadcasts on the radio and TV until you are told to evacuate.

Evacuation means to leave the area of actual or potential hazard. If an evacuation is ordered, follow the instructions of local officials regarding evacuation routes and the location of shelters. Take your emergency supply kit with you. Close car windows and air vents and turn off heater or air conditioner.

ARE YOU READY FOR A TORNADO?

Here's what you can do to prepare for such an emergency:

Prepare a Home Tornado Plan

Pick a place where family members could gather if a tornado is headed your way. It could be your basement or, if there is no basement, a centre hallway, bathroom, or closet on the lowest floor. Keep this place uncluttered.

If you are in a high-rise building, you may not have enough time to go to the lowest floor. Pick a place in a hallway in the centre of the building.

Assemble a Disaster Supplies Kit

Gather emergency supplies including: emergency medications, nonperishable food, a non-electric can opener, bottled water (at least three gallons per day per person), a battery-powered radio, flashlight, extra batteries, extra clothes, important documents, cash and credit cards, a first aid kit and other items for infants, elderly or disabled family members and pets.

Store supplies in a waterproof, easy-to-carry container, such as a plastic tub with handles.

Conduct periodic tornado drills so everyone remembers what to do when a tornado is approaching.

Stay tuned for warnings:

Listen to your local radio & TV stations for updated storm information.

• A tornado WATCH means a tornado is possible in your area.

• A tornado WARNING means a tornado has been sighted and may be headed for your area. Go to safety immediately.

When a tornado WATCH is issued:

• Listen to the local radio & TV stations for further updates.

Be alert to changing weather conditions. Blowing debris or the sound of an approaching tornado may alert you. Many people say

• It sounds like a freight train.

When a tornado WARNING is issued:

If you are inside, go to the safe place you picked to protect yourself from glass and other flying objects. The tornado may be approaching your area.

If you are outside, hurry to the basement of a nearby sturdy building or lie flat in a ditch or low-lying area if there is no shelter.

If you are in a car or mobile home, get out immediately and head for safety (as above). You'll want to get out of your car because it could be blown through the air or roll over you.

After the tornado passes:

- Watch out for fallen power lines and stay out of the damaged area.

- Listen to the radio for information and instructions.

- Use a flashlight to inspect your home for damage.

Do not use candles at any time.

Tornado victims need to take cover beneath the ground because of debris traveling at speeds as high as 200 mph.

VOLCANOES

People protect themselves from volcanic eruptions by monitoring earthquake activity as well as the raising of elevation from the buildup of gases below the earth's surface.

Here is a projected sequence of events for an eruption of a volcano like Mount Baker:

- Molten rock (also called magma or lava) oozes upward toward the surface of the earth in the volcano's existing vertical channels, but it cannot proceed to the surface because it is too thick, or the channels are blocked by older, solidified magma.

- The pressure of gases builds up behind the rising magma, usually causing many small earthquakes within the volcano.

- There is an explosive eruption and an eruption cloud, consisting of hot lava fragments, ash and volcanic gas proceeds outward, away from the volcano. This is called a pyroclastic flow. A pyroclastic flow affects only the immediate vicinity, but it destroys all life in its path and knocks down trees and structures, sometimes setting them afire.

- Ash is projected upward by explosions and heat. Heavier pieces fall to the earth short distances away, but lighter ash is carried high into the atmosphere, and is blown far away by the prevailing winds. Ash can fall in a thick layer, suffocating plants and animals, collapsing buildings, and choking machinery. Ash will eventually wash into rivers and streams and cause them to silt up and change their course.

- Fast-moving slurries of rock, mud and water (called lahars), that behave much like flowing concrete, follow river courses and destroy and bury bridges and buildings. Pyroclastic flows can melt snow and ice and become lahars.

Can volcano effects be prevented?

We cannot stop a volcano from erupting. We can, through awareness of what could happen, reduce some volcano effects. But, because volcanoes erupt so infrequently, the tendency is to accept or even ignore risks, or to assume that an eruption will not occur in a lifetime or more.

Awareness of volcano danger, emergency preparedness, and contingency planning to deal with the effects usually entails evacuations, emergency shelter for large numbers of people, extra strain on health care facilities, and recognition that the agricultural and fishing industries can be seriously (though perhaps only temporarily) damaged.

Fortunately, a volcano will often give a period of advance warning before serious effects result. Scientists and emergency managers on both sides of the Canada-USA border cooperate to assess the current situation of Mount Baker, and take steps to warn people and to implement emergency plans should danger be imminent.

Volcano victims are at risk of the lava flow and gases emitted from a large unpredictable eruption.

What you can do

Learn about the volcano hazards that could affect you at home, work or school.

Practice a home evacuation.

Plan what you and your family will do if you have to leave home.

Keep an emergency kit ready.

Participate in community emergency preparedness activities.

WILDFIRES

People protect themselves from wildfires by enforcing laws about campfires and wood clearing. Beetle-infested forests must often be purged because the devastation causes so much fuel to be available during events like lightning strikes.

Do not panic

Bushfires range from low intensity (low flames, little heat produced) and slow moving – which can be controlled safely – to high intensity infernos which race across the landscape and are virtually uncontrollable and deadly: what are you dealing with?

Assess the problem

i.   Personal safety – highest priority

ii.   Safety of others – high priority

iii.   Property – low priority (except a vehicle to escape in)

Options

i.   Fight? Is this a sensible strategy?Is any assistance available: who, where, & how soon?

ii   Flight? Where is it safe to go? When do I need to go?

iii   Telephone or radio a message to the local fire brigade or police.

Protective clothing

Cover up your skin (long sleeves, trousers), wear boots, a hat (ideally a helmet), gloves, a towel over your nose and mouth, eye protection (glasses or goggles) – everything to stop radiant heat or flames burning you and to minimize smoke inhalation: soak clothes in water if available. Ideally the fabrics should not be synthetics, which can melt, but cotton or wool.

Winter Storms

There are many kinds of winter storms. You probably already know those happening during the winter, like the snow storm and the ice storm.

Snow Storm

The temperature must be sufficiently cold for snow to fall and flakes to reach the ground. When winds blow heavily and you can not see before you, it means that a blizzard is blowing.

What to Do During a Blizzard

Go home! Many people lost themselves during major snow storms. Everything becomes white and it is hard to find your way back home.

Did You Know...?

- Have you ever examined a snowflake? They are all different. Even though most of them have six sides, you will never find two similar snowflakes!

- A big snowflake travels at 5 km/h.

- A blizzard buried a train in 1947 in Saskatchewan. The snowdrift was 1 km in length and 8 m in height.

Ice Storm

What is an ice storm? When it rains in winter, drops of water are very cold. When these cold drops hit something colder, they freeze and make a layer of ice. This layer is called glaze.

In Québec, glaze usually falls between December and March. In Montréal, there are approximately 13 days of freezing rain each year.

What to Do During an Ice Storm

Go home! Glaze makes streets, sidewalks and stairs very slippery. You could hurt yourself if you fall down. Be sure that your emergency kit is easily accessible.

Did You Know...?

- In Québec, during the January 1998 ice storm, 10 cm of glaze fell and 3 million persons experienced power failure.

Avalanche

An avalanche occurs when a mass of snow falls down a mountainside. That is because new snow (which is not wet) accumulates on a more heavy snow layer. Since the new snow layer is not compact, it could slide down toward the base of the mountain.

Avalanches could be triggered by loud noise (yells, firearm shots, etc.). Skiers, snowboarders and climbers could also cause an avalanche.

There are two ways to reduce avalanche risks: by installing snowsheds or by triggering controlled avalanches. These protection fences are made of rocks, soil and other materials. They prevent avalanches to fill trails used by humans.

Controlled avalanches cause the accumulated snow to fall down before it could trigger avalanches by itself. When nobody is in danger, explosives are used to produce a loud noise that causes an avalanche.

Avalanches could be seen every winter in mountainous regions of Québec. Fortunately, they do not often occur near inhabited areas.

What to Do During an Avalanche

During the winter, do not venture in the mountain without an adult companion.

If you walk in a trail after a snow fall, do not walk in the middle of it, but rather on the side. Avalanches always fall down in the middle of the path; thus, if you are on the side, the risks of being hit by the avalanche are lower.

After An Emergency

These are general instructions that apply to many emergencies but not every situation is the same. Try to stay calm.

- Check yourself and others for injuries. Give first aid to people who are injured or trapped. Take care of life-threatening situations first. Get help if necessary.

- Check on neighbours, especially the elderly or people with disabilities.

- Confine or secure pets.

- Use the battery-operated radio from your emergency kit to listen for information and instructions.

- Do not use the telephone except to report a life-threatening injury. Please leave the lines free for official use.

- If possible, put on sturdy shoes and protective clothing to help prevent injury from debris, especially broken glass.

- If you are inside, check the building for structural damage. If you suspect it is unsafe, leave and do not re-enter.

- Do not turn on light switches or light matches until you are sure that there aren't any gas leaks or flammable liquids spilled. Use a flashlight to check utilities.

- Do not shut off utilities unless they are damaged, leaking (a gas leak smells like rotten eggs) or if there is a fire. If you turn the gas off, don't turn it on again. That must be done by a qualified technician.

- If tap water is available, fill a bathtub and other containers in case the supply gets cut off.

- If there is no running water, remember that you may have water available in a hot water tank, toilet reservoir or in ice cube trays.

- Do not flush toilets if you suspect that sewer lines are broken.

- If you are in a high-rise building, do not use the elevator in case of power failure. If you are in an elevator, push every floor button and get out as soon as possible.

- Pick up your children from school or the pre-determined collection point.

- Stay away from damaged areas unless you are asked to help or are qualified to give assistance.

- Do not go near loose or dangling power lines. Downed power lines can cause fires and carry sufficient power to cause harm. Report them and any broken sewer and water mains to the authorities.

- Water supplies may be contaminated so purify your water.

- If the power has been off for several hours, check the food in the refrigerator and freezer in case it has spoiled.

CARING FOR THE INJURED

- Six basic first aid steps

- Bleeding

- Burns

- Injuries to muscles, bones and joints

- Exposure to chemical agents

- Reduce caregiver risks

These basic principles can be used in many situations. This information is no substitute, of course, for comprehensive first aid training or for proper medical care.

Six basic first aid steps

1. Survey the scene to make sure it is safe for you and others.

2. Check the victim for responsiveness. If the person does not respond, call for professional medical assistance (i.e. call 9-1-1 or other local emergency number).

3. Check and care for life-threatening problems, check the person's airway, breathing and

circulation, attend to severe bleeding and shock

4. When appropriate, check and care for additional problems such as burns and injuries to muscles, bones and joints.

5. Keep monitoring the person's condition for life-threatening problems while waiting for medical assistance to arrive.

6. Help the person rest in the most comfortable position and provide reassurance.

The following are some additional steps to take when providing care for common injuries:

Bleeding

Cover the wound with a dressing and apply direct pressure. If you do not suspect a broken bone, elevate the injured area above the level of the heart. Cover the dressing with a roller bandage to hold it in place. If the bleeding does not stop and blood soaks through the bandage, apply additional dressings and bandages without removing any of the blood-soaked dressings.

Provide care for shock. Encourage the person to lie down. Help the victim maintain normal body temperature.

Burns

Stop the burning by cooling the burn with large amounts of clean, cool water. Cover the burn with dry, clean, non-stick dressings or cloth. Do not break blisters.

Injuries to muscles, bones and joints

Rest the injured part. Avoid any movements that cause pain. Immobilize the injured part before moving the victim and giving additional care. Apply ice or a cold pack to control swelling and reduce pain. Elevate the injured area to help slow the flow of blood and reduce swelling.

Reduce caregiver risks

The chance of getting a disease while giving first aid is normally extremely low. To reduce the risk even further:

- Avoid direct contact with blood, other body fluids and wounds.

- Thoroughly wash your hands with soap and water immediately after giving care.

- Use protective equipment, such as disposable gloves and breathing barriers.

- Be aware of chemical/biological/radiological exposure risks.

COMMON EFFECTS OF AN EMERGENCY

While every situation is different, the following are some things you might expect during a disaster.

There may be a large number of casualties, the safety of buildings and houses may be compromised and rubble may block areas, making it dangerous or difficult to get out or walk around.

Health services in hospitals and mental health resources may become strained, maybe even overwhelmed. Know they are doing their very best under extraordinary circumstances. Health care facilities have emergency plans and might access additional resources - such as mobile hospitals - or enlist the support of medical staff and facilities from neighbouring communities, provinces or the Government of Canada.

Law enforcement from local, provincial and federal levels might be involved if the event was criminal in nature, such as a terrorist attack.

Extensive media coverage, high public anxiety and the social impact of the emergency could all continue for a prolonged period.

Workplaces and schools may be closed and there might be restrictions on local, domestic and international travel.

You and your family or household may have to evacuate an area following routes specifically designated to ensure your safety.

Clean-up and recovery operations could take many months.

Expect emotional reactions

People caught in a disaster often feel confused. You might not act like yourself for a while. You may tremble, feel numb, vomit or faint.

Immediately after the disaster, people often feel bewildered, shocked and relieved to be alive. These feelings and reactions are perfectly normal.

Many survivors sleep poorly, have no appetite, are angry with those around them or panic at the slightest hint of danger. Children might start thumb-sucking or bed-wetting. These feelings and reactions are perfectly normal too.

Getting Back on Track

Talk about your feelings. Discuss what's happened. Encourage your children to express their feelings. They may want to do this by drawing or playing instead of talking. Understand that their feelings are real.

Recognize that when you suffer a loss, you may grieve. (Yes, you can grieve the loss of a wedding photo or your grandfather's favourite ring.) You may feel apathetic or angry. You may not sleep or eat well. These are normal grief reactions. Give yourself and your family permission to grieve and time to heal.

Helping children cope

Children exposed to a disaster can experience a range of responses such as anxiety, fear, nervousness, stomach aches and loss of appetite. These are normal and temporary reactions to danger. Parents can help relieve their children's anxiety by taking their fears seriously, reassuring them and giving them additional attention and affection.

After a disaster, children are most afraid that the event will happen again, that someone will get hurt or injured, that they will be separated from the family or that they will be left alone. Comfort and reassure them. Tell them what you know about the situation. Be honest but gentle. Encourage them to talk about the disaster and to ask questions. Give them a real task to do -- something that helps get the family back on its feet. Keep them with you even if it seems easier to do things on your own. At a time like this, it's important for the whole family to stay together.

Television coverage of terrorist incidents and other tragedies -- and people's reaction to those events -- can be very upsetting, especially to children. Talk to your children about what is happening and how you and governments are keeping them safe.

We do not recommend permitting children to watch television reports that show images of the same incident over and over again. Young children often do not realize that it is repeated video footage and may think the event is happening again and again. Adults might also need to give themselves a break from watching disturbing footage. (Since listening to local radio and television reports will provide accurate information on what's happening and what actions you can take, try to take turns listening to the news with other adult members of your household.)

EPILOGUE

It is my hope that we are never required to respond to a disaster or emergency. However, it is my prayer that should we have to, not only will we draw from the information in this book but also from the unlimited resources of mercy and grace that can be found in our Lord and Saviour Jesus Christ.

Personal Notes:

Personal Notes:

_____

_____

_____

_____

_____

_____

_____

_____

_____

_____

_____

_____

_____

_____

_____

_____

_____

_____

_____

_____

_____

_____

_____

_____

_____